Scalars and Vectors

In physics, quantities can either be classified as scalars or vectors. The basis for classification is on whether the quantity can be fully described by its magnitude alone or whether it requires both the magnitude and direction for a complete description.

Let's quickly get through the 2 concepts.

1 The Concept of Scalars and Vectors

Vectors

> ### 1

A vector quantity is a quantity that has both magnitude and direction. Such quantities are not completely described unless their magnitude (or size) and direction are specified.

Examples of vectors include: displacement, velocity, acceleration, weight. force, momentum, electric field etc

Get it here!

> ### 2

If we say that a car travels at 100kmhr^{-1} due west, the magnitude (which is 100kmhr^{-1}) tells how fast it is travelling, and the direction of travel is west. Therefore the quantity just described is a vector. The only and sufficient condition for identifying a vector is that it should have a magnitude and a direction.

Scalars

| 3 |

Scalar quantities are those quantities that have only magnitudes, but not directions. Some examples of scalar quantities include: distance, speed, mass, and temperature.

The following statement, 'a car travels with a speed of 20kmhr^{-1}', does not tell us the direction of travel, and so represents a scalar.

Distinction between scalars and vectors

| 4 |

Scalars can be distinguished from vectors in the following way:

Scalars	Vectors
1. Have only magnitude	Have both magnitude and direction
2. Since scalars don't have direction, they are combined by ordinary algebraic methods	Vectors are combined by geometrical methods because of their directions. (This is illustrated shortly below).

2 Vector Representation (Graphical)

Plan 5

| 5 |

A vector is usually represented with a straight line which has an arrowhead indicating in the direction of the vector. The length of the line represents the magnitude of the vector, while the arrowhead points in the direction of the vector.

Get it in plan 6!

Vectors are arrow-headed lines

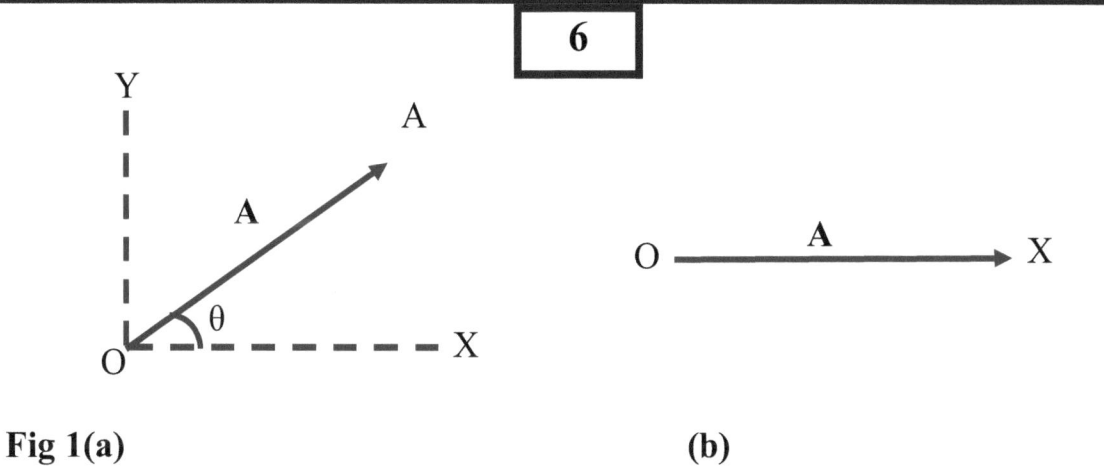

Fig 1(a)   **(b)**

In fig 1(a) above, a vector **A** is represented in magnitude by the length of the line OA and its direction is shown to be at angle θ to the X-axis. Fig 1(b) represents the same vector **A** but now in a direction along the X-axis.

3 Addition of vectors

Vectors acting along the same line

Vectors acting along the same line can be combined in a relatively simple way; we have two cases:
(i) when the vectors are in the same direction, and
(ii) when they are in opposite directions.

Case 1

Let's consider two forces of magnitudes P=30N and Q=40N, are acting on a body in the same direction, we want to know the magnitude and direction of the resultant force.

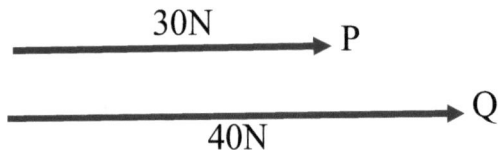

Fig 2

The magnitude of the resultant Force R is the algebraic sum of the two forces: That is, R=30N+40N =70N

And the direction of the resultant is the same as that of P and Q.

Case 2

Now let's consider the same two forces (P and Q) acting in opposite directions.

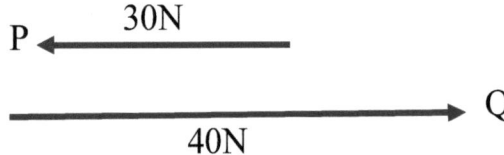

Fig 3

In this case the magnitude of the resultant is obtained by subtracting the smaller one from the bigger one, and the direction of the resultant is the same as that of the bigger one.

That is, Resultant R=Q–P = 40N–30N =10N in the direction of Q, which is the bigger force.

Note that if P and Q where equal in magnitude, then the resultant R will be equal to zero. i.e. R=0

Vectors inclined at an angle

| 10 |

In the case where the two forces (or vectors) are inclined to each other in such an angle that they don't act along the same line, then their resultant cannot be obtained as simply as we did above, we will require some special methods as will be illustrated below.

Note:
The sum of two vectors, **a** + **b** = **c,** written symbolically does not imply arithmetic addition unless **a** and **b** are in the same direction.

Plan 11

| 11 |

Suppose we have two vectors **a** and **b** represented as shown in fig 4(a) below

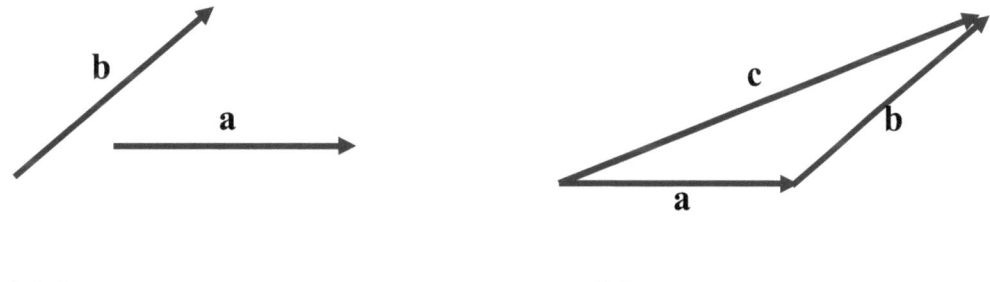

Fig 4 (a)   **(b)**

Two vectors can be added graphically as illustrated in fig 4(b) above by placing the tail of **b** to the head of **a** while maintaining the lengths and direction of both vectors. Their sum **c** can then be represented by the vector drawn from the tail of **a** to the head of **b**.

Thus, **c** = **a** + **b** is a symbolic representation of the addition operation.

With that we can now attempt the following JAMB question

12

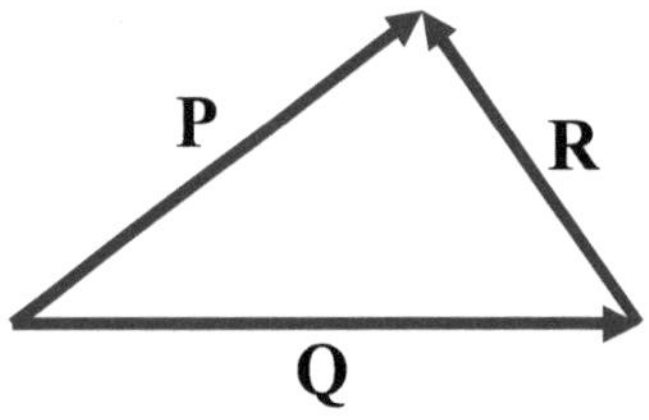

Fig 5

In the diagram above, **P**, **Q** and **R** are vectors. Which of the following options gives the correct relationship between the vectors?

(A) **P = Q + R** (B) **P = Q – R** (C) **P = R – Q** (D) **P + Q + R = 0**

Answer: If you chose option (A) then you are right.

From the diagram, we can see that vector **P** represents the addition of vectors **Q** and **R**; this is because the tail of **R** is on the head of **Q**, and **P** is drawn from the tail of **Q** to the head of **R**. So, **P = Q + R**.

If you got that right, let's now proceed with a numerical example

13

Let's find the sum of two vectors **A** = 5m <60^0, and **B** = 6m <130^0 interpreted as follows:

(i) Vector **A** has a magnitude of 5m and makes an angle of 60^0 with the horizontal, while

(ii) Vector **B** has a magnitude of 6m and makes an angle of 130^0 to the horizontal.

Let's do it!

14

First we represent vectors **A** and **B** as shown in fig 6 below

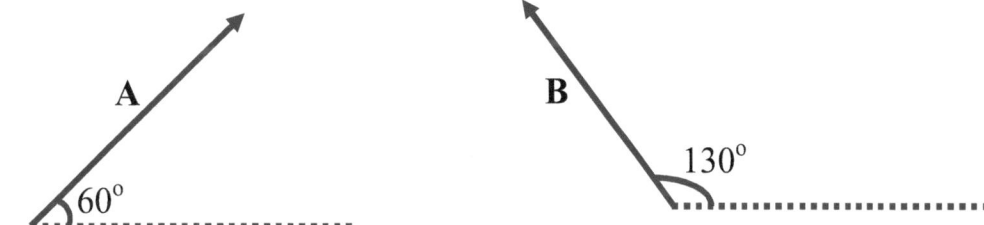

Fig 6

Next we combine the two vectors by placing the tail of **B** on the head of **A**, and then their sum **C** is drawn from the tail of **A** to the head of **B**.

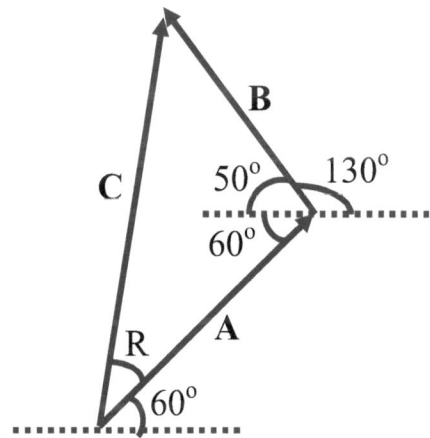

Fig 7

From the figure above, cosine rule gives the magnitude of the resultant **C** as:

$$\mathbf{C}^2 = \mathbf{A}^2 + \mathbf{B}^2 - 2\mathbf{AB}\cos(110)$$

$$\mathbf{C}^2 = 25 + 36 - 2(5)(6)\cos(110)$$

$$\mathbf{C}^2 = 81.52$$

$$\mathbf{C} = \sqrt{81.52} = 9.02m$$

15

Using the sine rule, we have

$$\frac{\sin R}{B} = \frac{\sin 110}{C}$$

$$\sin R = \frac{B \sin 110}{C}$$

$$\sin R = \frac{6(0.91)}{9.02}$$

$$\sin R = 0.6263$$

$$R = \sin^{-1} 0.6263 = 39^0$$

From our figure, we can see that the angle **C** makes with the horizontal is $(60 + R)^0 = (60 + 39)^0 = 99^0$

Therefore the vector **C** is given in magnitude and direction as $9.02m < 99^0$

If that is clear, let's proceed!

4 The Parallelogram Law of Vector Addition

Vectors acting at a point

16

The parallelogram law of vector addition is used to find the resultant of two vectors acting at the same point.

The law states that if two vectors are represented in magnitude and direction by the adjacent sides of a parallelogram, then their resultant can be represented in magnitude and direction by the diagonal of the parallelogram drawn from the point of intersection of the two vectors.

17

If two vectors, say **P** and **Q**, are represented by the adjacent sides of a parallelogram as illustrated in fig 8(a) below, then we can complete the parallelogram and represent their resultant **R** with the diagonal of the parallelogram drawn from the point of intersection of **P** and **Q** as shown in fig 8(b).

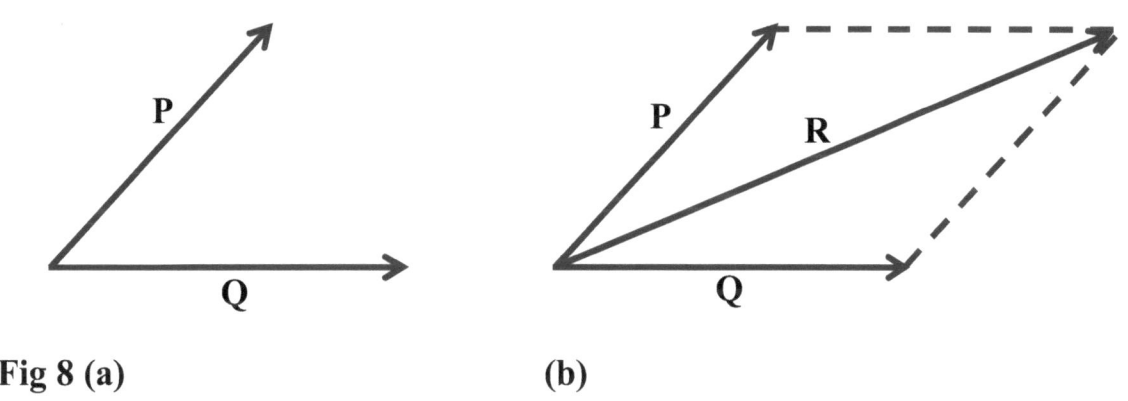

Fig 8 (a)   **(b)**

We can use the method of either
    (i)   Scale drawing, or
    (ii)   Trigonometric and geometric formulae
to obtain the magnitude and direction of **R**.

By Scale drawing

18

  The following steps will be taken to get the magnitude and direction of the resultant by scale drawing.

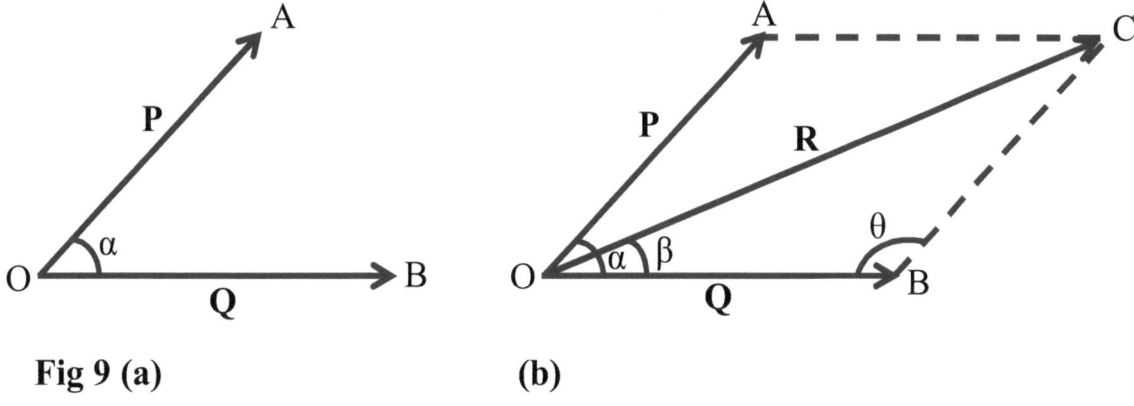

Fig 9 (a)   **(b)**

1. First a suitable scale is chosen so as to accommodate the lengths of the vectors to be drawn on the drawing paper.

2. Using the chosen scale, lines OA and OB are then drawn to represent vectors **P** and **Q** respectively, at the appropriate angle between them, say α, as shown in fig 9(a) above.

3. The parallelogram is completed by drawing AC parallel and equal to OB and BC parallel and equal to OA. (Fig 9(b)).

4. The line OC is then drawn to represent the resultant vector **R** as shown in fig 9(b). The magnitude of **R** is obtained by measuring the length of OC and converting it using the chosen scale. The direction is that of the arrow pointing from O to C, which can be measured from the diagram as the angle β which the resultant makes with **Q**.

Using trigonometric and geometric formulae

Here, we need just to sketch the vectors, and not necessarily to draw them to scale.

We can obtain the magnitude of **R** by cosine rule:
$$OC^2 = BC^2 + OB^2 - 2(BC)(OB)\cos(\theta)$$
That is, $R^2 = P^2 + Q^2 - 2PQ\cos(\theta)$
And so $R^2 = P^2 + Q^2 + 2PQ\cos(180-\theta)$ [Because $\cos(\theta) = -\cos(180-\theta)$]

Also, since the adjacent angles of any parallelogram add up to 180, we have that α+θ=180 or α=180-θ, and so we can replace 180-θ with α to get;

$$R^2 = P^2 + Q^2 + 2PQ\cos(\alpha)$$

Therefore, in general…

20

If two vectors P and Q are inclined at angle α to each other, then the magnitude of their resultant R is given by:

$$R^2 = P^2 + Q^2 + 2PQ\cos(\alpha) \qquad\qquad (1)$$

Next is the direction

21

We can get the direction of the resultant vector R by using the sine rule:

$$\frac{\sin(\beta)}{BC} = \frac{\sin(\theta)}{OC}$$

That is, $\dfrac{\sin(\beta)}{P} = \dfrac{\sin(\theta)}{R}$

$$\sin(\beta) = \frac{P\sin(\theta)}{R}$$

And so $\beta = \sin^{-1}\left(\dfrac{P\sin(\theta)}{R}\right)$

This is the angle which the resultant makes with vector Q.

If that is clear, let's consider a numeric problem

22

Find the resultant displacement of the following 2 vectors; 7km N30^0 E and 10km East.

Solution:

First, we need to represent the given information in diagram.

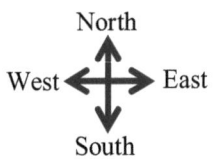

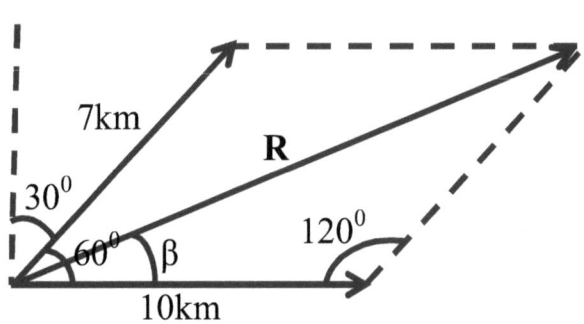

Fig 10

To get the magnitude, we use the cosine rule (in the form of Equation 1);
$R^2 = 7^2 + 10^2 + 2\,(7)(10)\,Cos\,60^0$
$R^2 = 49 + 100 + 140\,(0.5)$
$R^2 = 219$
$R = \sqrt{219} \qquad = 14.80km$

And to get the direction, we can use the sine rule;
$$\frac{\sin\beta}{7} = \frac{\sin 120}{R}$$

$$\sin\beta = \frac{7\sin 120}{R} \qquad = \frac{7\,(0.8660)}{14.80} \qquad = 0.4096$$

$\therefore \beta = \sin^{-1}(0.4096) \quad = 24°$

Therefore, the resultant displacement is: 14.8km at an angle of $24°$ with the 10km vector.

5 Resolution of Vectors

The idea!

Any vector **A** lying between two perpendicular directions (say x and y) has some contributions it is making to both directions.

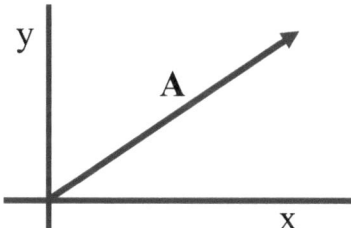

Fig 11

The contribution of **A** to the x direction is called the component of **A** in the x direction, while the contribution of **A** to the y direction is called the component of **A** in the y direction. **To resolve a vector in any direction means to find its component in that direction.**

How to resolve vectors

Let's illustrate how to resolve vectors by resolving vector A in the x and y directions. Let's call the component of **A** in the x direction A_x and its contribution in the y direction A_y as shown below.

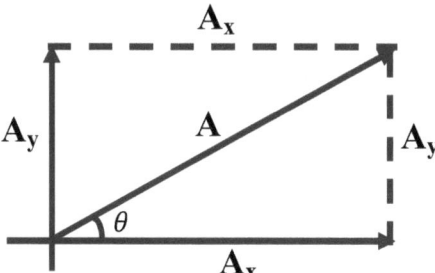

Fig 12

Now from trigonometry, we can see that

$$\cos\theta = \frac{A_x}{A} \quad \text{and that} \quad \sin\theta = \frac{A_y}{A}$$

Therefore
$$A_x = A\cos\theta \qquad\qquad 2(a)$$

and
$$A_y = A\sin\theta \qquad\qquad 2(b)$$

A short clue to resolving vectors

| 25 |

In general, if a vector of magnitude, A, is lying between two perpendicular directions (x and y) such that the vector makes an angle of θ with either of the directions (say x), then the component of vector A in this x direction is **A cosθ** while its component in the other direction (now y) is **A sinθ**

Please! Please!! Please!!! Note this fallacy:
Some clues say that resolving to the horizontal (or x direction) is always **A cosθ** while resolving to the y direction is always **A sinθ**. This is really very wrong!

If vector **A** makes an angle of θ with the vertical (or y direction), then the component of vector **A** in the vertical (or y) direction will be **A cosθ**, while in the horizontal (or x) direction it will be **A sinθ.**

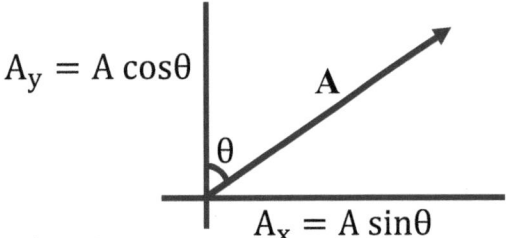

$A_y = A\cos\theta$

A

θ

$A_x = A\sin\theta$

Fig 13

So, it all depends on the direction with which the vector makes the angle.

The correct clue once again says:
It is [A cosθ] for the direction with which vector A makes the angle θ, and [A sinθ] for the other direction.

26

A body of weight WN rests on a smooth plane inclined at an angle Θ^0 to the horizontal. What is the resolved part of the weight in Newtons along the plane?

(A) $W\sin\Theta$ (B) $W\cos\Theta$ (C) $W\sec\Theta$ (D) $W\tan\Theta$

Any idea? You should make a sketch first, then the picture gets clearer.

Solution

27

Let's make a sketch of the diagram.

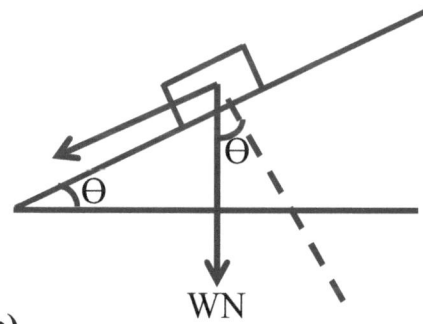

Fig 14(a)

Note that we have drawn the weight WN to be vertically downwards from the center of the body. We have also drawn 2 directions that are perpendicular to each other (that is, along the plane and perpendicular to the plane). Finally, geometry shows that the angle of the inclined plane Θ is equal to the angle which the weight makes with the perpendicular to the plane.

Now, since the angle Θ is between the weight W and the perpendicular to the plane, we resolve in the direction perpendicular to the plane to get $W\cos\Theta$, and in the direction along the plane to get $W\sin\Theta$.

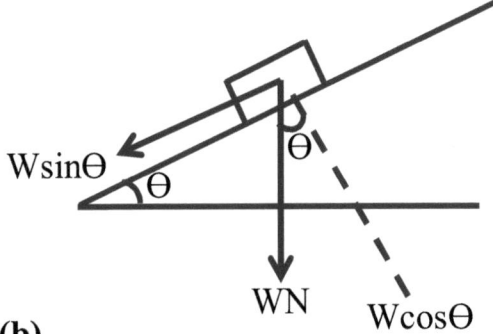

Fig 14(b)

Therefore, the resolved part of the weight in Newtons along the plane is $W\sin\Theta$. Option **A** is correct.

28

Two forces of magnitudes 7N and 3N act at right angles to each other. The angle Θ between the resultant and the 7N force is given by

(A) $\cos\Theta = \dfrac{3}{7}$ (B) $\sin\Theta = \dfrac{3}{7}$ (C) $\tan\Theta = \dfrac{3}{7}$ (D) $\cot\Theta = \dfrac{3}{7}$

Solution!

29

A sketch of the diagram looks like this:

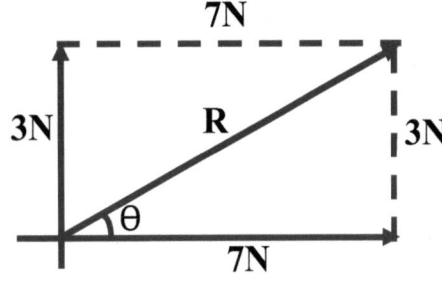

Fig 15

By trigonometric ratios;

$\tan \Theta = \dfrac{3}{7}$

So, option **C** is correct.

Simple, isn't it!

6 Resolution of more than two vectors

Plan 30!

30

When we have more than two vectors to resolve, the resultant of the system of vectors (say **a**, **b**, **c**, **d**, - - - - - - - -) can be determined a follows.

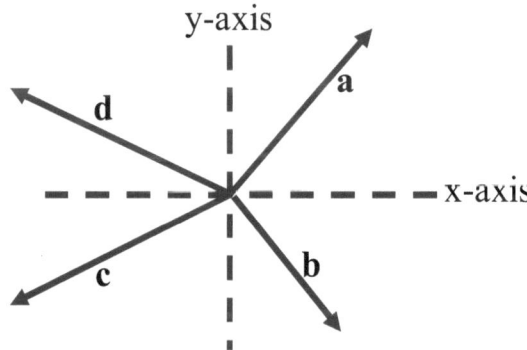

y-axis

x-axis

Fig 16

Step 1: Find the component of each vector along x-axis

Step 2: Find the component of each vector along y-axis

Step 3: Find the arithmetic sum of these resolved parts along the x-axis. [Let's call it sum(x)].

Step 4: Also find the arithmetic sum of the resolved parts along y-axis. [Let's call it sum(y)].

Step 5: The resultant is then given by: $R = \sqrt{(sum(x)^2) + (sum(y)^2)}$

| 31 |

$$\text{sum}(x) = \mathbf{a_x} + \mathbf{b_x} + \mathbf{c_x} + \mathbf{d_x} - \ - \ - \ -$$
$$\text{sum}(y) = \mathbf{a_y} + \mathbf{b_y} + \mathbf{c_y} + \mathbf{d_y} - \ - \ - \ -$$

The magnitude of the resultant vector is:
$$R = \sqrt{(\text{sum}(x)^2) + (\text{sum}(y)^2)}$$

And the angle the resultant vector makes with the x-axis is:
$$\propto = \tan^{-1}\left(\frac{\text{sum}(y)}{\text{sum}(x)}\right)$$

where $\mathbf{a_x}$, $\mathbf{b_x}$, $\mathbf{c_x}$, $\mathbf{d_x}$, - - - - respectively represent the components of vectors $\mathbf{a}$, $\mathbf{b}$, $\mathbf{c}$, $\mathbf{d}$, - - - - along the x-axis, and $\mathbf{a_y}$, $\mathbf{b_y}$, $\mathbf{c_y}$, $\mathbf{d_y}$, - - - - represent those along the y-axis.

Let's illustrate numerically with this JAMB question

| 32 |

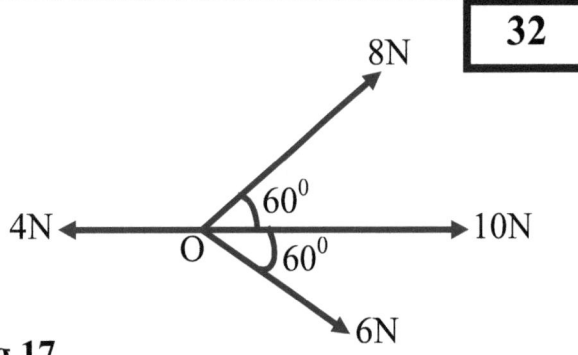

Fig 17

The diagram above shows forces 4N, 6N, 10N and 8N which act at a point O in the directions indicated. The net horizontal force is

(A)13N   (B)17N   (C) $\sqrt{3}N$ (D) $7\sqrt{3}N$

Let's do it!

The net horizontal force is the net force along the x-axis, therefore we need just to resolve the forces along the x-axis, and not necessarily along the y-axis. This is done below: (remember it is 'cos' when the angle is between the vector we are resolving and the axis to which we are resolving it, otherwise it is 'sin'):

for the 8N = $8\cos60^0$ N in the positive x-axis $\quad = 8 \times 0.5$ N $= 4$N
for the 6N = $6\cos60^0$N in the positive x-axis $\quad = 6 \times 0.5$ N $= 3$N
for the 4N = 4N already in the negative x-axis $\qquad\qquad\qquad = $ -4N
and for the 10N = 10N already in the positive x-axis $\qquad\qquad = $ 10N

Therefore, the net horizontal force is
sum(x) = 4 + 3 + (-4) + 10 = 13N. Option **A** is correct.

Now try the following

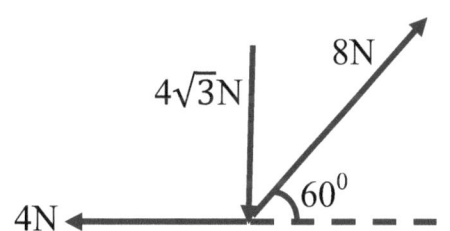

Fig 18

Obtain the resultant of the system of forces shown the diagram above.

Attempt it on your own before looking at the solution below.

Solution:
Resolving all the forces to the horizontal axis we get:
for the 8N = $8\cos60^0$ N along the +ve x-axis $\quad = (8 \times 0.5)$ N $\quad = 4$N
for the 4N = 4N already in the -ve x-axis $\qquad\qquad\qquad = $ -4N

The $4\sqrt{3}N$ vector has no component on the horizontal axis because it is acting exactly along the y-axis.

Therefore sum(x) = 4 + (-4) = 0

Similarly we look for sum(y) by resolving the forces to the vertical axis:

for the 8N = 8sin60^0 N along the +ve y-axis $= (8 \times \frac{\sqrt{3}}{2}) \, N$ $= 4\sqrt{3}N$

for the $4\sqrt{3}N = 4\sqrt{3}N$ along the -ve y-axis $= -4\sqrt{3}N$

And so sum(y) = $4\sqrt{3}N + (-4\sqrt{3}N) = 0$

The resultant vector is therefore
$R = \sqrt{(sum(x)^2) + (sum(y)^2)}$ $= \sqrt{0^2 + 0^2}$ $= 0$

7 Relative Velocity

The idea!

| 35 |

Relative velocity describes the velocity of a body relative to another. It is a purely vectorial concept since velocity itself is a vector.
What we do here is very much similar to what we have done earlier in the section on addition of vectors.

Get it clear

| 36 |

If two bodies A and B are travelling in a straight line, the velocity of A relative to B is found by adding the velocity of B reversed to the velocity of A.
Let us take an example to illustrate this point.

Here is one.

37

A car travelling on a straight road at 10km/h passes a bus going in the same direction at 5km/h, the velocity of the car relative to the bus is given by?

Going by our explanation in plan 36 above, the relative velocity of the car will be the velocity of the bus in reverse direction plus the velocity of the car.

That is, -5 + 10 = 5km/h

The real meaning of this is that someone in the car notices that the car is moving at 5km/h relative to the bus. Relative to the bus, the car appears to be moving slower than it is actually doing because the bus is also moving in the same direction.

If the bus were to be moving at exactly the same speed as the car, then the car won't appear to be moving at all relative to the bus; the relative velocity will be 0.

What if they were both moving in opposite directions

38

Suppose the car (moving at 10km/h) and the bus (moving at 5km/h) were moving in opposite directions. What will be the velocity of the car relative to the bus?

The velocity of the car relative to bus in this case is (5 +10) = 15km/h.

Since the bus is moving in an opposite direction to the car, the car will appear to be moving faster than it is actually doing relative to the bus.

39

Another valid question is: what if the two velocities are inclined at an angle (other than 0^0 and 180^0) to each other?

Here is a practical example:
Suppose the car is travelling at 10km/h due south of a place O, and the bus is travelling at 5km/h due east of the same place O. What will be the velocity of the car relative to the bus?

the steps

40

(i) represent the question in a diagram:

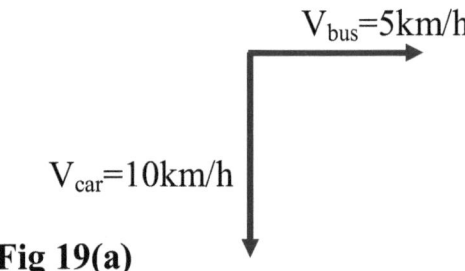

Fig 19(a)

(ii) reverse the direction of the bus since we are looking for the velocity of the car relative to it. That is:

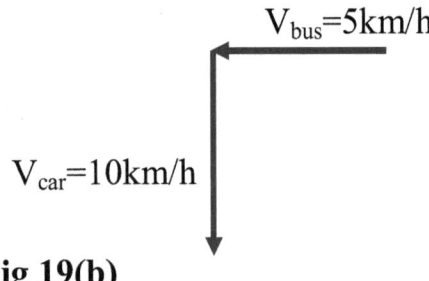

Fig 19(b)

(iii) use the laws of vector addition (discussed in plan 11) to draw the resultant **R** of the two vectors:

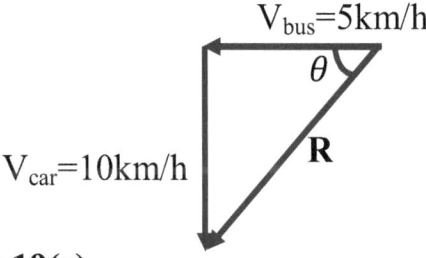

Fig 19(c)

R represents the relative velocity of the car relative to the bus.

Its magnitude is:

$$R = \sqrt{10^2 + 5^2} \quad = 5\sqrt{5} \text{ km/h}$$

And the angle it makes with the west direction is:

$$\theta = \tan^{-1}\left(\frac{10}{5}\right) \quad = 63.4°$$

1 A ball is moving at 18ms^{-1} in a direction inclined at 60^0 to the horizontal. The horizontal component of its velocity is:
(A) $9\sqrt{3}ms^{-1}$ (B) 6ms^{-1} (C) $6\sqrt{3}ms^{-1}$ (D) 9ms^{-1}
(JAMB)

2 Which list contains only scalar quantities?
A acceleration, displacement, mass
B acceleration, distance, speed
C displacement, mass, velocity
D distance, mass, speed
(Cambridge)

3 A body on the ground is acted on by a force of 10N at a point as show in the diagram below. What force is needed to stop the body from moving eastwards?
(A) 5N in the direction of East
(B) 5N in the direction of West
(C) $5\sqrt{3}$N in the direction of West
(D) 10N in the southwest direction
(JAMB)

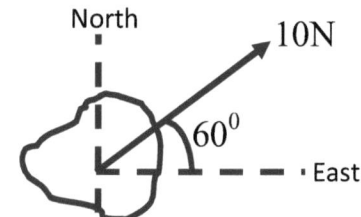

4 The resultant of two forces 12N and 5N is 13N. What is the angle between the two forces?
(A) 0^0 (B) 45^0 (C) 90^0 (D) 180^0
(JAMB)

5 An air craft attempts to fly due North at 1000km/h. If the wind blows against it from east to west at 60km/h^{-1}, its resultant velocity is
(A) 117km/h, N31^{0}E
(B) 127km/h, N31^{0}E
(C) 117km/h, N31^{0}W
(D) 127km/h, N31^{0}W
(JAMB)

6 When two forces are combined, the size of the resultant depends on the angle between the two forces.
Which of the following cannot be the magnitude of the resultant when forces

of magnitude 3 N and 4 N are combined?
(A) 1 N   (B) 3 N   (C) 7 N   (D) 8 N
(Cambridge)

7 Two forces F1 and F2 act on an object O in the directions shown.

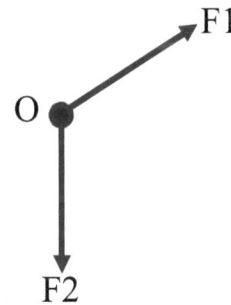

What is the direction of the resultant force?

(A)   (B)   (C)   (D)

(Cambridge)

8 Which diagram correctly shows the addition of a 4N and 3N force?

(A) 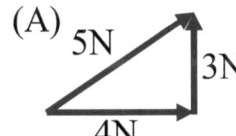   (B) 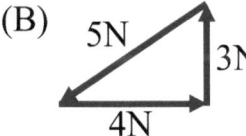   (C)    (D)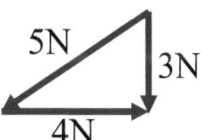

(Cambridge)

9 Two forces act at right angles at a point O as shown.

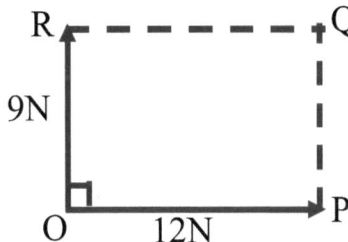

What is the resultant of the forces?

	Magnitude	Direction
(A)	15N	OQ
(B)	15N	PR
(C)	21N	OQ
(D)	21N	PR

(Cambridge)

10 A man walks 3 miles north then turns right and walks 4 miles east. The resultant displacement is:
(A) 1 mile SW
(B) 7 miles NE
(C) 5 miles NE
(D) 5 miles E
(http://www.edinformatics.com/math_science/vectors_scalars.htm)

11 Find the magnitudes of Q and W if the forces below are in equilibrium.

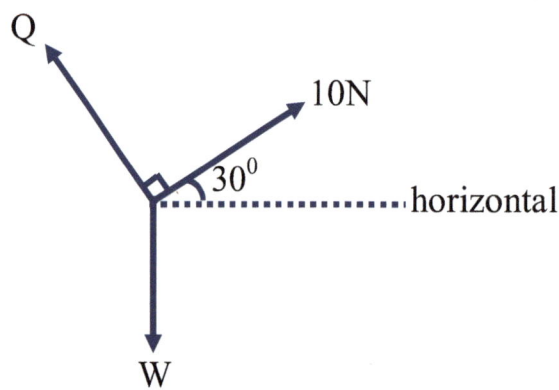

12 Two vectors V_1 and V_2 are inclined to each other at an angle g,
(a) determine the magnitude of their resultant expressed in terms of V_1, V_2, and g.
(b) what angle does g represent when the vectors are (i) in opposite directions? (ii) in the same direction? (iii) perpendicular to each other? (iv) parallel to each other?
(c) at what value of g is the magnitude of the resultant equal to V_1+V_2?

Solutions to Exercises

1 D

2 D

3 B

4 C

5 C

6 D

7 D

8 A

9 A

10 C

11 $Q = 10\sqrt{3}$ N, $W = 20$ N

12 (a) Resultant $= \sqrt{V_1{}^2 + V_2{}^2 + 2V_1V_2 \cos(g)}$

(b)(i) 180^0 (ii) 0^0 (iii) 90^0 (iv) 0^0 or 180^0

(c) 0^0